Beyond Pyramid Power

The Science of the Cosmos II

G. Patrick Flanagan, Ph.D.

PHI SCIENCES PRESS
Cottonwood, AZ 86326

Copyright © 1975 by G. Patrick Flanagan Ph.D. All rights reserved, foreign and domestic. No part of this book may be reproduced in any form, or by any means, mechanical or electronic, without permission of the copyright holder.

ISBN-10: 1530859158
ISBN-13: 978-1530859153

Library of Congress Catalog Card Number: 73-86022

Printed in the United States of America
Cover design by Russell Chong

To My Eve

GLOSSARY

FREQUENCY: The number of repetitions of a periodic process per unit of time. Usually denoted in cycles per second.

HARMONIC or OVERTONE: A vibration whose frequency is an integral multiple of a fundamental frequency.

INTERFERENCE: The process in which two or more vibrations combine to reinforce or cancel each other, the amplitude of the resulting wave being equal to the sum of the amplitude of the combining waves.

STANDING WAVE: A wave disturbance which is not progressive. Standing waves result from the super-position of two vibrations traveling in opposite directions, having identical frequencies and amplitudes. The wave disturbance is maximal at certain points occurring periodically, called loops or anti-nodes, and it is zero at points in between called nodes. The distance between any two nodes is a wavelength.

WAVELENGTH: The distance between two successive points in a wave.

NODE: A point, line, or region in a standing wave where there is little or no vibration. A point of stillness in a sea of vibration.

G. PATRICK FLANAGAN

Since I wrote my two previous publications on pyramid energy in 1971 (*The Pyramid and its Relationship to Biocosmic Energy* and *Pyramid Power*) I have had over four years of additional experience with these unusual energy fields. The results of this research have never been made public.

I will reveal for the first time, the results of these discoveries, and details of the construction of the Pyramid Energy Plates.

Most of these discoveries were made in a six month period following my trip to the Great Pyramid of Giza in September, 1974. On that trip, I culminated years of research into unusual energies by spending the night in the King's Chamber of the Pyramid of Cheops. Full details of that exciting night will be given in a forthcoming book.

I can reveal that I made contact with a powerful force that has changed my life in drastic ways. Some of my latest findings may tend to contradict a few of my earlier statements on the subject. This is the result of an increased understanding of the mechanics of these energy fields.

I have succeeded, since returning from Egypt, in harnessing these elusive energies in practical ways. Energy fields have been produced and transmitted over long distances and detected at a remote receiver. In other words, a new communications technology has been discovered.

Pyramid energy has been taken from the realm of the mystical toy, and projected into the infancy of a new technology that could literally change the living condition of man on this planet. I call these new discoveries *Tensor Field Devices*. A few of these

BEYOND PYRAMID POWER

developments are: Tensor Field Voice Communicator; Tensor Field Propulsion; Tensor Field Electro-Catalytic Seed Treatment, which raises the vigor level of seeds from 60% to 80%; and Tensor Field Water Treatment Systems. Additional devices will be described toward the end of this paper.

The mysterious properties of the pyramid shape are now pretty well known to the general public. My previous articles on the subject, dozens of national magazine articles, and over 600 hours of radio and television interviews have continuously created great interest.

The shape of the pyramid has been experimentally found to have unusual effects on organic media. It was found to mummify meat and other foods, preserving them without decay; improve the flavor of foods and cigarettes; age wine and other drinks; increase the life-span of small animals; improve the growth rate of plants; improve or enhance the meditative state; and keep razor blades from becoming dull!

There has been some controversy on this subject as test results are not always consistent. We have found that many variables affect the pyramid's performance: alignment to the earth's magnetic fields; electric fields; other objects in the vicinity; and the experimenter's mind can sometimes influence the results. Sorting out and isolating these variables has been the subject of my research over the years. As my own results were over 90% consistent, I had enough evidence to keep my interest alive.

Even N.A.S.A. has realized the importance of the pyramid shape as evidenced by a new patent application showing the use of pyramids and cones as sources of electric power.

G. PATRICK FLANAGAN

Beyond Pyramid Power

June 1973

B73-10185

NASA TECH BRIEF
Goddard Space Flight Center

NASA Tech Briefs announce new technology derived from the U.S. space program. They are issued to encourage commercial application. Tech Briefs are available on a subscription basis from the National Technical Information Service, Springfield, Virginia 22151. Requests for individual copies or questions relating to the Tech Brief program may be directed to the Technology Utilization Office, NASA, Code KT, Washington, D.C. 20546.

Proposed Electromagnetic Wave Energy Converter

The problem:
Electric power can, theoretically, be transmitted in several ways. But the only method that is currently practical on a large scale is through metallic conductors. As evidenced by the "universal" distribution and use of electricity, this method has worked well enough in the past. However, expanding demands for power are requiring the development of power sources that are new, and perhaps more "universal" in the literal sense of the word. Currently under consideration, for instance, are orbiting power stations which would beam power down to the earth from miles out in space. Such a system would require development of an efficient power converter for transforming electromagnetic radiation to useful electrical power. In addition, such a converter in a more advanced form could potentially be used to convert energy from the largest available power supply, the sun, permitting widespread terrestrial application of solar-electric converters. Such converters for centralized or dispersed use, as on homes, would capitalize on our only inexhaustible nonpolluting energy source, the sun.

The solution:
A proposed device converts incident electromagnetic wave energy into electric power through an array of insulated absorber elements responsive to the electric field of impinging electromagnetic radiation. This device could also serve as a solar energy converter that is potentially less expensive and fragile than solar cells, yet substantially more efficient.

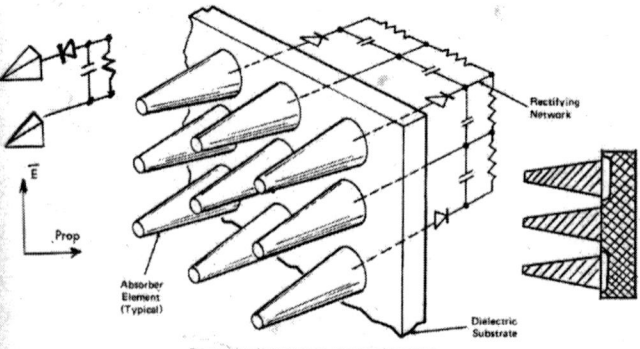

Electromagnetic wave energy converter for vertical, horizontal, or randomly polarized waves

(continued overleaf)

This document was prepared under the sponsorship of the National Aeronautics and Space Administration. Neither the United States Government nor any person acting on behalf of the United States Government assumes any liability resulting from the use of the information contained in this document, or warrants that such use will be free from privately owned rights.

BEYOND PYRAMID POWER

A portion of the NASA Tech Brief on the *Electro-Magnetic Wave Converter* and a few drawings from the patent application are included for the reader.

This device is designed to receive randomly polarized microwave signals coming from the Universe. I had a small company in Glendale, California, that manufactured and sold pyramids for people who did not want to build their own. I am no longer connected with that enterprise.

I have found a number of shapes more efficient than the pyramid. The pyramid has served its purpose as a beacon from thousands of years ago; pointing the way for twentieth century man to realize a lost technology.

It was thought that a man by the name of Bovis was the first person in recent times to discover that the pyramid shape could mummify foods. Bovis went to the Great Pyramid in the 1930's and found the mummified remains small desert animals in the King's Chamber. When he returned to France, he built a three foot base pyramid of wood and put a dead cat inside. In a few days, he had a mummified cat with no evidence of decay. This pointed the way to other experiments with different perishables and he found he could preserve all kinds of foods in this manner.

G. PATRICK FLANAGAN

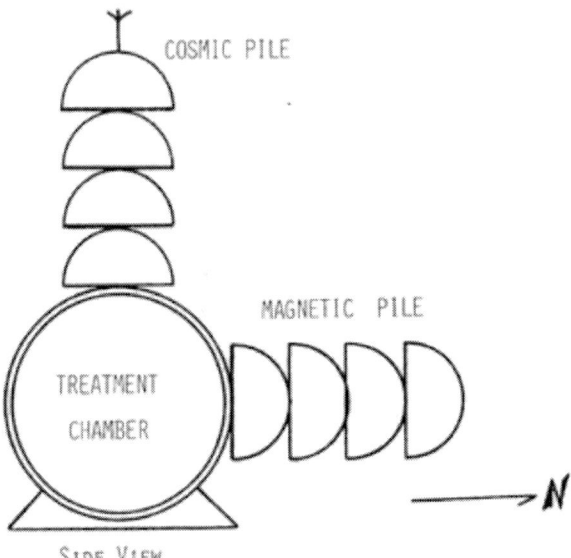

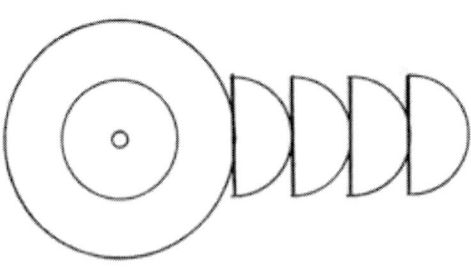

The Belizal Cosmic Accumulator, the "Bombe C-30". Used for treating herbs and mummifying foods in the central sphere. This device was developed in 1926.

BEYOND PYRAMID POWER

I found a book by another Frenchman, a water dowser by the name of Belizal. After years in Egypt, Belizal wrote a book, published in 1926, entitled *Ondes De Forme* (Waves of Form). In this book, he discusses his research into ancient Egyptian healing techniques. He claims the Egyptians used a number of shapes including the pyramid as healing energy sources. Belizal called the energy emanating from these devices the *Vert Negatif*, Negative Green Ray. Belizal said the Vert Negatif could be used for healing, treating foods and herbs, and for preserving foods without decay. He said that the Egyptians found the sphere more powerful than the pyramid.

The Vert Negatif, prana, mana, chi, ki, odic force, orgone, tumo, animal magnetism and other names have been given to these energies. Recently, the Russians have called it Bio-plasma, and I called it Bio-Cosmic or etheric energy. The descriptions of these energies are all similar. I have now given these forces another name, one more accurate for reasons I will describe. I now call these energies Tensor Fields.

ENERGY HAS SHAPE

In the physical universe, all things have structure. This means all physical principles can be modeled. All previous pyramid researchers believed that the pyramid shape had energy. The truth is the reverse of that statement; ENERGY HAS SHAPE! Modern Science went slightly astray when it stopped modelling the Universe and started describing it only in abstract mathematical terms. This has created a great chasm in understanding.

G. PATRICK FLANAGAN

We can state in abstract mathematical terms the characteristics of radio waves. These terms are very hard to visualize; however, modern radio antennas attest to the fact that energy has shape. There are thousands of radio antennas designed to match the shape of electro-magnetic energy. There are three dimensional and planar or flat arrays designed for this purpose.

Dr. Harold Saxon Burr, a professor at Yale University, wrote a book entitled, *Fields of Life*. In this book; Dr. Burr discusses several decades of research into the electric fields surrounding living organisms. A flower seed was found to have an electric field that had the exact shape the flower would have been when it was mature. A frog's egg had an electric field that was shaped like a frog! His conclusion was that the purpose of these fields is to control the formation and growth of the organism into its mature state, and to maintain the integrity of the cell structure of the mature organism.

The electric field is an invisible blueprint directing new cells to their proper places. Although this field can have electric effects, I have found this field to be another kind of energy. I call it the Bio-Tensor Field.

The ancient Science of acupuncture has caused the Western world to take notice of the Bio-Tensor Field. The basis of acupuncture is that there is another circulatory system in the body; a system of energy fields.

Blockages of this energy result in pain and sickness, and such blockages are the cause of aging and death. The needles redirect these fields, releasing the blockage and resulting in energy balance. The Western world is so busy concentrating on material aspects of existence, that it missed the invisible lines of force that control the physical. We are too busy treating the effect rather than the cause. Energy has shape; it controls the physical. What controls the energy? Under certain conditions, it is possible for the mind to control these energy fields. All forms of spiritual healing and miraculous regression of illness are caused by thought controlling energy. It is well known in physics that the higher the frequency of energy, the more physical power it contains. The energies we are dealing with are so high in frequency, their frequencies are almost impossible to differentiate.

Two forces are found in mechanical energy—the force of pressure, or compression; and the force of tension. R. Buckminster Fuller is perhaps the world's greatest expert on the interplay of these forces. Where we have pressure, we have tension. These forces are almost inseparable. In building construction, compression members have a maximum size ratio beyond which they become weak. The finest steel known to man has a maximum length to width ratio of 33 to 1. Steel tension cables used in bridge construction remain strong no matter how long they become.

The force of tension can be much more efficient than the force of compression. A tension wave can be communicated over longer distances than a compression wave. Remember the old Dixie cup intercom we made as children? It consisted of a piece of string tied to the ends of two small thin plastic cups. When tension was placed on the string and one of the cups was spoken into, it responded by varying the tension in the string in resonance to the speech vibration. The diaphragm on the cup at the other end vibrated in unison and duplicated the sound on the other end. Relax the string and there is no communication. Replace the string with a long wooden rod or a solid iron rod and no communication takes place. Nature uses combinations of tension and compression in most economical ways. Nature always maximizes efficiency. There is abundance for all in nature, but she wastes nothing.

In occult study, there is an old Hermetic saying: "As above, so below; as below, so above". This is an ancient truth that indicates the interplay of the physical and the metaphysical. The law of nature is uniform from one dimension to another; from one density to another density.

In the study of electro-magnetic energy, all energies are known to exert pressure in the direction of propagation. This effect has been proposed by NASA as a possible means of vehicle propulsion on long space voyages.

A great sail would be erected in space, and the pressure of the sun's rays would propel the vehicle and its cargo on its journey. The pressure of Laser light has been dramatically demonstrated in a recent experiment in which a laser was able to suspend a small glass sphere against the force of gravity. This is much the way a balloon is suspended on an air current and captivated above an electric fan. The pressure of the laser was actually overcoming the force of gravity!

It is a wonder that modern science has neglected to realize that pressure is always associated with tension. As above, so below; as below, so above.

BEYOND PYRAMID POWER

Dr. N. A. Kozyrev, the most respected astrophysicist in the world, has written a paper entitled: Possibility in the Experimental Study of the Properties of Time. Dr. Kozyrev discusses his theory of time as an energy. He demonstrates experimental evidence in support of his hypothesis. His description of the energy of time is similar to descriptions of pyramid or tensor energy. In one experiment, he stretches a rubber band and his delicate instruments detect a new kind of energy emanating between the cause and the effect ends of the band. His instruments were activated even though he isolated the rubber band by three feet thicknesses of copper and lead!

No shield has been found to isolate the force of gravity. It is unaffected by such things as copper and lead. Electro-magnetic energy is impeded by copper and lead. Dr. Kozyrev suggests that gravity is a time field. Gravity is indeed a Tensor Field; none of us are oblivious to its constant pull.

Wallace Minto of Sarasota, Florida has discovered what he calls hydronic and plasmonic signals being emanated from all species of fish. He found that fish are able to communicate over great distances under water, distances impossible to traverse using the world's most powerful radio transmitter. He found there are a number of nerves on the side of the fish that are designed to detect these radiations.

The Neurophone, a hearing aid I developed as a teenager used the same kind of energy to transmit sound to nerve centers located on the surface of the skin of the human body. Completely deaf people were able to hear with the device. We later discovered the "nerves" detecting these signals were actually acupuncture points.

G. PATRICK FLANAGAN

I had discovered a way to generate acoustic tensor signals by using brute force. I now have a development that generates these signals without resorting to power wasting brute force radio techniques. The Neurophone had the unique characteristic in that it was able to transmit a signal through a shield made of lead and copper.

A tensor field is always associated with electro-magnetic waves. The electro-magnetic vectors balance the tensor vectors, making the tensor field hard to detect.

There are certain techniques that separate these vectors and intensify the tensor field while minimizing the electromagnetic vector.

I have said in the past that pyramid energy is the energy that holds matter together. After reading of Dr. Kozyrev's experiment with the rubber band, I decided to test this with my own tensor field measuring instruments. I believed Dr. Kozyrev was detecting the stress he was placing on the molecular bonds of the rubber band. As I stretched the rubber band near my sensor, the meter registered a powerful response.

I then tried another experiment. I took a rectangular plate of copper and placed it in a vise. As I applied pressure to the sides, the center started to bow into a concave-convex surface. As increasing pressure was applied, the plate registered a powerful response on my meter. I then wrapped the ends of the plate with fiberglass strapping tape so I could remove it from the vise without releasing the tension. I then placed a small metal plate in the tensor field and connected the plate to a cassette tape

machine. I connected headphones to my tensor field meter so I could hear the signal across the meter. As I played a cassette tape into the tensor field, I could hear the sound coming from the tensor director! I could hear the sound coming from the tensor detector!
I had discovered the tensor field communications device!

I have since succeeded in transmitting this signal over great distances. It is impossible to detect the signal with common radio equipment. The tensor beam from this device is extremely directional.

Tensor fields have powerful effects on liquids. It is possible that the molecular bonds of liquids absorb and store tensor energy as water stores heat when it is heated. This would explain Water anomalies noted by Michael Gauquelin in his book: *Cosmic Clocks*. Scientists around the world are growing frustrated by water that changes its boiling point and freezing point, electrical conductivity, and the rate at which certain chemical reactions occur.

Under certain conditions, metals absorb these fields and re-emit them at a slow rate. In 1971, I built the first model of the pattern now known as a pyramid energy generator. The first one was made of six inch base aluminum pyramids arranged in a 3 x 3 grid. The harmonious pattern of the pyramids reminded me of the microwave antenna patterns I have seen, all spaced at certain wavelength intervals. I postulated that the harmonious grid arrangement placed the pyramid peaks at an even multiple of wavelengths apart. Having previously realized the effect of the earth's high voltage ionospheric field on the energy of the Great Pyramid, I duplicated this field by placing a metal plate under the pyramids, and another one several inches above the peaks. I then connected a high voltage DC power supply to the plates. I created a voltage gradient of over 1(X), (IX) volts per meter. A field T 1,000 times that of the Earth.

BEYOND PYRAMID POWER

TENSOR FIELD PLATE: When this copper plate was bent into a convex-concave surface under great stress, it became an intense source of Tensor Energy.

PYRAMID ENERGY PLATE CHARGER: Pyramid grid in conjunction with a high voltage source to generate intense Tensor Fields.

G. PATRICK FLANAGAN

When small metal plates were placed in this pyramid field, they absorbed a charge of tensor energy. These would then retain the charge for several months. These pyramid plates are not to be confused with similar plates produced by other researchers. One such device, the purple energy plate was being made and sold in California ten years before I made my pyramid generator. It was charged by a machine called the micro-magnetron developed by Dr. Jensen of Arizona. It consisted of a glass tube with electrodes at either end. In the center were two magnets. When a 1000 volt battery was attached to the electrodes, its inventor claimed it emitted an energy that also charged plates.

The pyramid energy plates were manufactured and sold with a money back guarantee and out of thousands sold, only a few were ever returned. They can be recharged by placing them in a pyramid overnight.

I have since discovered a more powerful device based on spheres and hemispheres. It is called a sphere generator.

A Kirlian Photograph was made of the pyramid generator by putting the generator in a vacuum. When high voltage was applied from outside the vacuum, the energy fields became luminous. A drawing of the original Pyramid Energy Generator was published in my Doctoral Dissertation in 1971.

R. Buckminster Fuller has written a fascinating book entitled: *Synergetics*, published by MacMillan and Company. The book gives considerable detail on the geometry of nature. Whereas Dr. Fuller has gone into these geometries as architecture, I have learned how to measure and use the energy fields to which these

BEYOND PYRAMID POWER

structures are resonant.

In *Pyramid Power*, not to be confused with the paperback book with the same title published a full year after my book came out, I proposed a structure for the ether or fluid matrix filling the Universe. This elementary thesis was presented to explain certain phenomena. As a result of additional years of meditation on the subject, I have an addition to that theory.

Edward R. Dewey and Og Mandino wrote a book entitled: *Cycles*. In this book, the authors discuss years of research funded by Govemment grants. This was research into possible correlations between cyclic events.

One of the startling discoveries revealed in this book is that there is no such thing as a random cycle! All cycles of the same duration have the same phase alignment. Things that appear random are merely random because the clock controlling these events has not been found. For example, the 8 year cycles of whiting abundance, sweet potato production, pig iron prices, price of butter, price of sugar, cigarette production, Goodyear Stock started, peaked, and reached lows in perfect phase alignment. This book proves beyond doubt that there is some kind of universal clock or power controlling things.

Universal Energy Structure

Consider an energy clock somewhere in the Universe; we could call it the Divine One. This clock is the source of all energy. It would have to consist of twelve sources of energy in order to fill the Universe with harmonic order. These sources are radiating discrete spherical packets of energy. As these spherical vibrations expand, they interfere with each other producing stable patterns of nodes and anti-nodes. In order for the pattern to be stable and self-regenerating, the patterns must intersect at exactly the same angles and have exactly the same distance between all points of intersection. In other words, the pattern would multiply in precise duplicating structural alignment, filling space as a web of energy interferences.

There is only one pattern that meets these requirements, a pattern of three sided (minus base) pyramids known as tetrahedrons, and double four sided pyramids known as octahedrons. This is the famous Octet-Truss pattern of Dr. Fuller.

Each node point acts as a new spherical source of energy and forms interference patterns of vectors radiating. In the Science of Moiré Patterns, we can represent a cross section of a spherical source of waves by a series of concentric circles. As these patterns are brought within each other's influence a standing wave of interference is generated. A few combinations of such patterns are illustrated in the diagrams. If these patterns are visualized three dimensionally, a structural web is generated.

These interferences are lines of force between energy events. These lines look solid, and they look as though they are standing still. If the concentric circles are made so close together that the eye cannot differentiate, we can see only the interferences and commonly mistake the interference for the source. THIS IS THE GREAT ILLUSION OF THE UNIVERSE. This is the illusion spoken of by mystics and philosophers for thousands of years.

BEYOND PYRAMID POWER

Examples of spherical wave interference patterns. The concentric rings represent a cross section of energy radiating from a spherical source. The interference patterns form the illusion of solid lines between the fields.

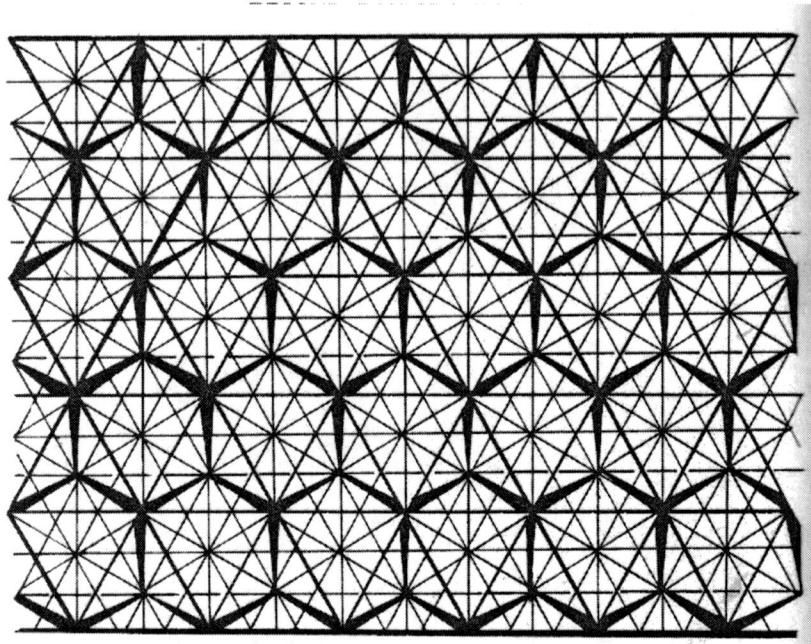

INTERFERENCE PATTERNS from multiple spherical sources fill space with an omni-triangulated pattern of stable standing wave patterns.

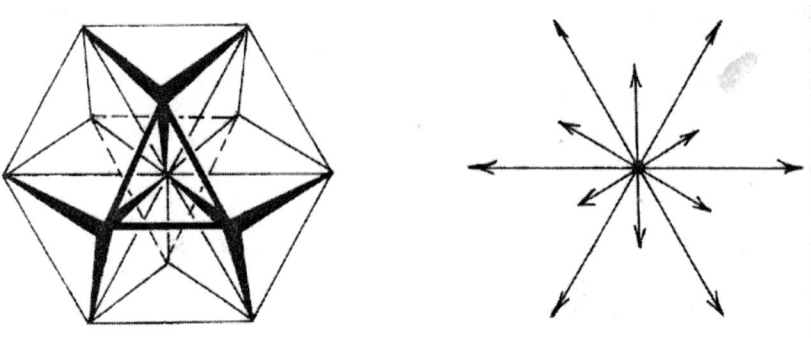

Every nodal point is a spherical source of twelve radiating vectors. These are equiangular and equal in length.

BEYOND PYRAMID POWER

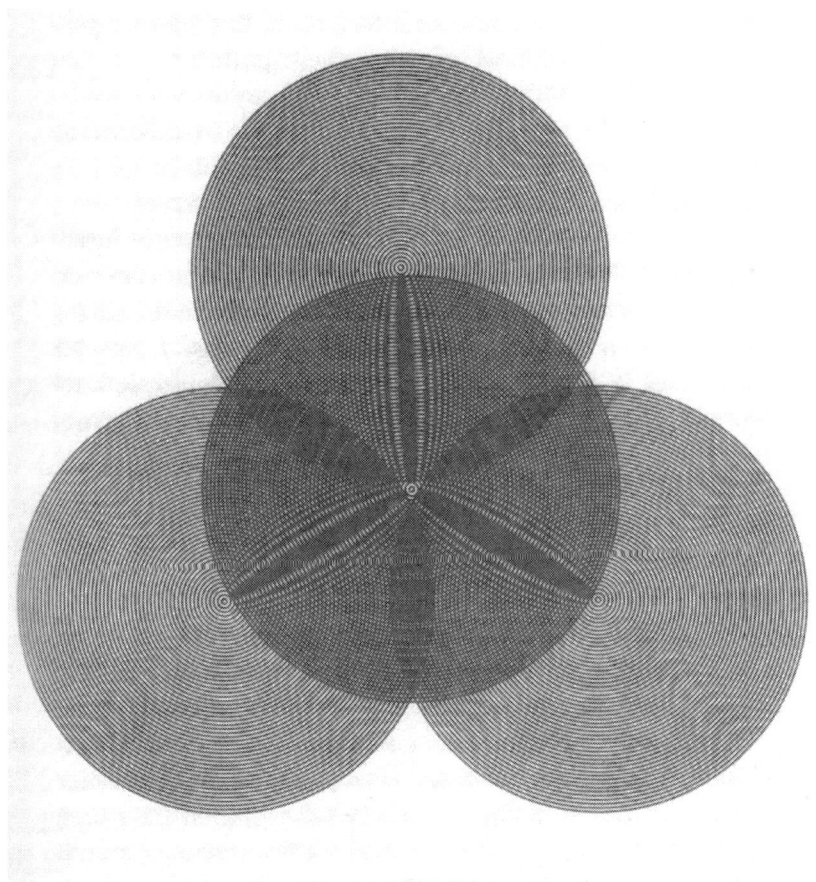

As the frequency of the spherical source gets higher, the wavelets are harder to see, but the interference patterns become more obvious.

Our universe is composed of interference patterns of energy events. We only see the interference and cannot perceive the invisible field creating these patterns.

These patterns generate corresponding structures at harmonic intervals of one another. These interpenetrate with common nodes where intersecting lines of wave propagation of higher frequencies coincide with interferences of lower overtone harmonics. At these common node points where harmonic intervals begin and end together, there can be a transmission of energy from one dimension to another. If we consider these multiple harmonics as varying degrees of density, we have a structure for a multi-dimensional Universe. Awareness of any dimensional plane is dependent on one's ability to differentiate specific energy events. The interference patterns generated in this manner form the structure of all energy. The expanding spherical field being pressor energy interferences, and the standing wave patterns being the Tensor Field events.

It can be said that the Universe is a controlled explosion of interfering energy events. Pyramids and other structures tune to the gross patterns that match the shape of the structure, in the same way radio antennas match the shape of the pressor radio waves. The power of energy everywhere in the Universe cannot be denied when we see NASA applying for patents on pyramid array to respond to these events.

By generating tensor beams in specific angular patterns, it should be possible to generate artificial resonant imploding tensor node points. This would be a miniature "black hole".

I believe the Atomic bomb's release of enormous energy is caused when the tensor binding field of an unstable radio isotope is weakened by an impinging pattern of spherical pressor fields. The release of the tensor controlling force allows the matter to be released in its full explosive radiance.

It is possible that the interconnecting tensor fields of the Sun and planetary systems align at certain times to create multidimensional node points. These node points would correspond with certain geometrical patterns to create twelve such points on Earth. These patterns would occur on a cyclic basis. This could be an explanation of the Devil's Triangle in Bermuda. At certain times, the alignment of forces would create a window to other density levels. Actual records of disappearances in these areas indicate they are indeed cyclic.

I am planning an expedition into the Bermuda triangle with my measuring equipment to test this theory. We should be able to detect these multidimensional points with the tensor measuring equipment.

All matter is a controlled explosion; radial spherical pressure outward, tensional force restraining. If the bonds of tension are released, this explosion expands at the speed of light squared. Total energy released is $E=MC^2$.

Our ability to differentiate minute energy events is the basis of what we call time. When modern science discovered the basic blocks called the proton, the neutron, and the electron, these were considered to be the smallest particles of matter. As resolving power became greater, the electron resembled a cloud of particles. Finer and finer particles are being discovered almost daily. It is becoming obvious that matter is made of energy. The physical world is simply a pattern of energy interferences; a great illusion of solidarity and stillness in a web of motional energy fields.

G. PATRICK FLANAGAN

Although the above may seem esoteric, it is a brief outline of an energy structure that makes possible all known energy phenomena. There may be flaws, but I believe it is a highly accurate description.

The science of modern man has concentrated on pressor energy phenomena. Architecture has been in the realm of compressional structures. Our buildings are built of heavy objects and are held together by weight instead of a balance of tension and compression. The biggest example is the Great Pyramid. The reason it has lasted all these centuries is because it is massive. The blocks are so heavy they cannot go anywhere. The glue in between the blocks does not hold it together, it simply keeps the stones from slipping out of alignment while under construction.

Nature builds by balancing tensional and compressional energy. A tree is a perfect example. Water is incompressible. The tension of the cellulose structure of the tree against the incompressibility of water in its capillary tubes gives the tree its strength to withstand the elements. A tree can raise a column of water hundreds of feet into the air. Man with his most powerful vacuum pump can pull water a fraction of that distance. The tree uses tension to lift its water; biologists call it shoot tension. In the study of tensor energy phenomena, I have found dozens of shapes that are receptive to tensor fields.

A few of these are illustrated in this paper. I have found after careful research that the most efficient of passive devices are the sphere and hemisphere. Combinations of tetrahedrons and octahedrons form different states in between.

The Sensor: Rings of Fire

The Sensor is a planar Tensor Field Device. It is the most powerful planar device I have found. It represents a cross section of waves from a spherical energy source. It is analogous to some forms of planar—refractive microwave antennas. It is composed of a number of 24K gold-plated copper rings on a dielectric substrate. It does all the tensor field phenomena performed by the three dimensional shapes such as the pyramid and hemisphere.

The Ion Generator

The effects of negative ions are well known. It is believed that the prana mentioned in Hindu texts is the negative voltage in the air. It has been shown that ions have powerful regulating effects on the endocrine glands. In nature, these ions cleanse the air of pollutants. Russian research has shown that animals cannot live in environments where ions are removed from the air. I have developed an ion machine for experimental research in tensor effects. The unit generates ions without Ozone, and doubles as a power source for pyramid plate charging and creation of electrical tensor fields. It is especially needed indoors to activate the pyramid.

G. PATRICK FLANAGAN

Tensor: Electro-Catalytic Seed Stimulant

Research into techniques of releasing tensor field energies from mineral substances has made possible this exciting discovery. I have found a way to dissociate minerals to release tensor fields for increasing the growth and health of plants. This new farming method is being successfully tested in the United States and several foreign countries. It has resulted in a 50% increase in crop yield.

It can be used in a number of ways: as a seed treatment, combined with electronic seed stimulation; irrigation drip; foliar spray, a method of feeding by mist spray. The roots of crops treated in these ways are resistant to herbicides, pesticides, pests, and frost bite.

The Tensor Athletic Pad

The athletic pad is a voltage stimulated tensor field device. A twelve volt pulsating field is used in a tensor array to create a powerful tensor field. It is called an athletic pad as a number of athletes are using it to prevent sore muscles. Soreness is apparently an energy congestion of acupuncture energy fields. The pad provides a source of tensor energy to break these field patterns and return them to normal.

Sphere Generators

Sphere generators are passive receptors of tensor fields. A number of spherical devices have been constructed, some after Belizal and some of my own design. One of these devices was so powerful it ionized the air in the environment. The photographs are of some designs. The planar sphere generator is similar in use to the Pyramid Energy Generator. It is more powerful than the pyramid generator. When used in conjunction with a voltage field, it will also charge energy plates.

G. PATRICK FLANAGAN

THE SENSOR: A planar Tensor Field Receptor

THE ION GENERATOR: Used for generation of Tensor Fields and the charging of Pyramid Energy Plates

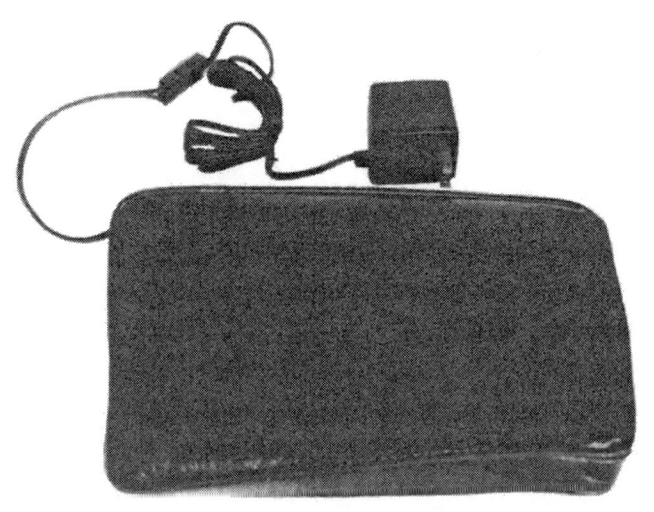

ATHLETIC PAD: A source of tensor field waves.

G. PATRICK FLANAGAN

Bearing Generator

The steel bearing generator in the photograph is another example of spherical field source. It is constructed of styrofoam and a number of stainless steel ball bearings placed in a triangular Grid. This resembles the pattern of the snowflake.

Tubular Pyramid Antenna

Many researchers have had considerable results meditating in a tubular pyramid antenna. These pyramid structures are made of hollow aluminum tubes one inch in diameter. This is not unlike some of the VHF television antenna currently on the public market. The tubular pyramid if properly constructed is one of the most powerful of the pyramid designs.

Cone of Fire

A cone is also a powerful tensor shape. If we start with a tetrahedron, and consider it the simplest pyramid structure, we can expand it by adding triangular faces. The second phase is the correspondence to the Great Pyramid.

As we increase the frequency of faces, the structure resembles a smooth surfaced cone. At this point, the cone-pyramid is no longer sensitive to alignment to the earth's magnetic field. A number of cone devices have been built and tested. They are more powerful than the pyramid.

BEYOND PYRAMID POWER

The Indian tipi is an example of a cone type receiver. It may explain some of the high energy states the Indians were known to possess.

Mental Control of Tensor Fields

A Sufi Master from Istanbul, Hassan Shushud, taught me a Caucasian Yoga breathing exercise in the summer of 1974. When this exercise is performed properly, it has a very powerful effect in the Bio-Tensor field surrounding the Body.
Tantric Yoga sexual exercise techniques previously unknown in the West also have powerful effects on the BioTensor fields, charging the Tensor field with a great deal of energy. By performing Caucasian Yoga exercises these energies may be directed by the mind. They can produce motional effects on matter. These fields, thought directed, easily register on my tensor field detectors. The techniques for increasing and directing these fields mentally cannot be described in this work.

A seminar is being prepared to make these techniques available. When we realize that thought can be directed to produce physical energy effects, and that the power of these thoughts can be increased to create more powerful effects, the significance of these statements becomes obvious. ENERGY SHAPES THE PHYSICAL, THE MIND CAN SHAPE AND DIRECT THE ENERGY.

Not only are these practices beneficial to the practitioner, they can be beneficial to mankind if directed by group thought processes. The student can measure his progress with tensor field detectors.

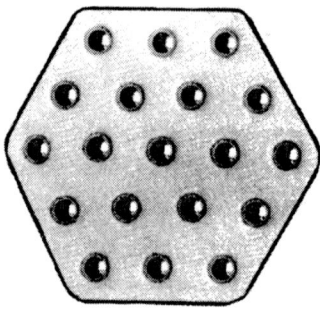

STEEL BEARING GENERATOR: Pattern of steel ball bearings in Styrofoam has been used as a microwave antenna, and has been found efficient as a Tensor antenna. In microwave technology it is used as a refractor array.

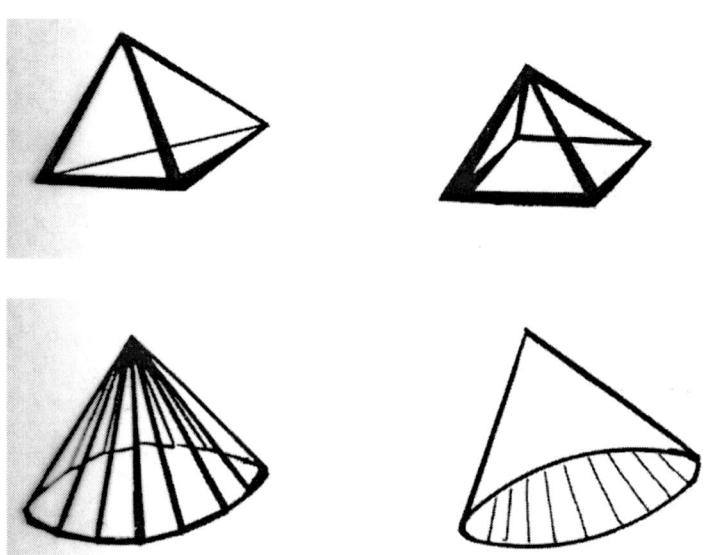

BEYOND PYRAMID POWER

As the number of triangles on the side of a Pyramid are increased, it becomes a cone. The cone-pyramid is not sensitive to alignment to the earth's magnetic field.

FLANAGAN RESEARCH FOUNDATION

The amount of information on Tensor Field Effects is encyclopedic and it is an enormous task to get it together. This field is still in its infancy although I have demonstrated practical uses now being applied for the benefit of man. I am forming a corporation called INNERGY INC., and I need help.

Those who would like to help are urged to write:

>Dr. G. Patrick Flanagan
>271 S Airpark Rd,
>Cottonwood, AZ 86326
>Help@PhiSciences.com

G. PATRICK FLANAGAN

APPENDIX

PYRAMIDS,
CONES
AND BIOCOSMIC ENERGY

BEYOND PYRAMID POWER

PYRAMIDS, CONES, AND BIOCOSMIC ENERGY

Summary of Background

The idea that the pyramid shape has mysterious powers dates back beyond recorded history. When we were in Egypt in September, 1974, the Arab guides and Bedouin chieftains had many tales of the mysterious powers possessed by the Great Pyramid. They told all kinds of tales of mysterious healings, divine revelations, and forces surrounding the pyramids.

Various religious cults have expounded the powers, claiming the pyramid is a sacred chamber for initiation into higher consciousness. The Egyptian Book of the Dead contains numerous references to the pyramid as a chamber of initiation. Manly P. Hall wrote that an initiate would enter as a man and leave three days later as a god, having had the secrets of the Universe revealed in the interim.

In the 1920's, a French dowser by the name of Belizal wrote numerous books and pamphlets on the power of pyramids, cones, spheres, and hemispheres. He wrote that these shapes contained mysterious properties, the *vert negatif* . . . or negative green ray, and that this ray was capable of mummifying food without decay, and of charging water and herbs with healing powers.

Later on, in the 1930's another French dowser named Bovis wrote articles on the power of the pyramid. It was Bovis who visited the Great Pyramid and found the mummified remains of small desert animals in the King's Chamber of the Great Pyramid. He found that the small animals were perfectly mummified with no decay. When he returned to France, he built a number of small three foot base pyramids from plywood. He succeeded in preserving cats, dogs, fish, brain tissue, liver tissue, vegetables, fruits and other foodstuffs. He published these findings in various pamphlets and books. Over the next several decades, interest in pyramid energy slowly grew and others around the world picked up on the experiments of Belizal and Bovis.

In California, Verne Cameron, another famous dowser built and experimented with cones, pyramids and other shapes. He published his results in various newsletters. It was Cameron who placed a piece of meat infested with maggots under a cardboard pyramid. He made a small window in the side to view the progress of the experiment. Within a few hours, the maggots left the meat and starved to death. It is well known that maggots feed on decaying tissue. It was obvious from this experiment that the pyramid stopped and possibly reversed the decay process to the extent that the maggots were no longer interested in the meat.

Similar experiments by myself and others indicated that a properly constructed and aligned pyramid could prevent decay and preserve a number of other items. In the early 1950's, a Czechoslovakian radio engineer named Karel Drbal started experimenting with pyramid models. Drbal approached the pyramid from an electromagnetic viewpoint, believing the pyramid was a microwave resonator. He discovered that the pyramid model would also sharpen razor blades so a person can shave as many as 200 times from a single razor blade. Drbal was so impressed by this discovery that he applied for, and received a patent on the pyramid razor blade sharpener. In 1959, he received patent number 91304, entitled Cheops Pyramid Razor Blade Sharpener.

In his argument to the patent examiner, he proved that a pyramid shape could focus enough energy from the universe to affect organic substances and speed the rate of rotation of water dipoles and thus speed dehydration.

Ostrander and Schroeder in *Psychic Discoveries Behind the Iron Curtain*, enlightened the general public on the powers of the pyramid.

From the above background, it can be easily shown that the subject of pyramid energy is by no means a new subject.

AUTHOR'S BACKGROUND

My own background in pyramid research is covered in the book, *Pyramid Power*, first published by DeVorss and Company, Santa Monica and recently reprinted by Phi Sciences Press in Cottonwood, Arizona. A summary of that is included here. I have been interested in unusual energy fields since I was eight years old. At that time, I read several books on the subject of Yoga, and became intensely interested in the forces described in Eastern literature, especially the finer forces in the air and in living organisms; such as prana, the etheric energy in the air; tumo, the Tibetan word for life force in the body; Ki, the Japanese word; Chi, the Chinese word for life force discussed in acupuncture; Odic force, the word for etheric energy given by Reichenbach; Orgone, the word of Reich; Mana, the Hawaiian word given by the Kahunas; and other words by other researchers. I call it biocosmic energy.

G. PATRICK FLANAGAN

I started a study of esoteric sciences as a result of this interest. By the time I was sixteen, I was experimenting with mental control of these energies and had devised various way of measuring the results of my experiments on my own body.

My development of the Neurophone at the age of fourteen was instrumental in my research into the subject. The Neurophone won world-wide publicity as it was an electronic sixth sense, a special type of radio transmitter that bypassed the eighth cranial nerve and transmitted sound information directly to the brain centers. This device made it possible for some totally deaf people to hear for the first time in their lives.

The Neurophone was unique in that no one could figure out how it worked. According to previous knowledge of the human nervous system, the Neurophone could not work. But we were daily demonstrating electronic telepathy. We later learned that the Neurophone was actually transmitting sound directly to the brain by means of the acupuncture energy meridians.

Another interesting side note was that the Neurophone bypassed the normal sensory channels and enabled simultaneous communication to both hemispheres of the brain and actually increased the IQ of the user as a result of the increased neural efficiency.

When I read of the mysterious forces present in the shape of the pyramid, I began an intense series of experiments in an effort to unravel the secrets locked therein. As a result of my previous experiments with the Neurophone, acupuncture, and electric fields surrounding animals and plants, I used all the tools available to me in this quest.

BEYOND PYRAMID POWER

I like to follow word meanings back to source in my studies. As a result of tracing the word pyramid back, my appetite for knowledge was whetted even more. I found that the word pyramid means "fire in the middle".

Pyro = fire, heat, or light. Amid: in the middle. It seems that the ancients knew more than we had given them credit for. I naturally duplicated the experiments of Belizal and Bovis. Indeed, under proper conditions the food left in the pyramid mummified or dehydrated with little or no sign of decay. Control foods placed in equal volume cubical boxes decayed quite normally. I also found that a pyramid has a limit to its capacity. When the item to be treated gets beyond a certain volume ratio to the volume of the pyramid, the pyramid no longer has the capacity to affect the substance. I will get into this ratio in the experiment section.

When I got into pyramid research, there were only two things known about the pyramid: 1. The pyramid mysteriously mummifies food. 2. The pyramid mysteriously sharpens razor blades.

As a result of my food experiments, I wondered what the preserved food tasted like. So I tasted it and found the items tasted better than they did when they were raw. In another series of experiments, I found that the taste of foods was affected within a few moments in most cases.

Bitter foods lost bitterness, sweet foods tasted sweeter, carbonated beverages went flat, cigarettes were milder, wine tasted smooth as if it had been aged for years, vodka tasted as though it had been charcoal filtered hundreds of times, fruit was more lively, even meat was reported to be more tender.

Further experiments carried out by my readers have shown that many small test animals such as mice and some microorganisms live up to nine times longer in a pyramid, tropical fish retain their superior genes after ten or twenty generations when they normally degrade drastically after a few generations. Mice raised in a pyramid were smarter than their parents. Meditation in a pyramid results in alpha brain waves double or triple the normal amplitude, indicating a higher state of consciousness.

Sex in a pyramid is enhanced far above normal sex in a cubical room. This last item has resulted in more pyramid experimentation than all the other items mentioned.

Plants raised in pyramids grow faster and healthier than plants grown in cubical boxes. This last item has resulted in a number of new and exciting innovations in the agricultural area, including a new and superior method of treating seeds, This new method, entitled the ECSS, or Electro Catalytic Seed Stimulant is now in use in over 5,000 acres of crop land from Canada to Texas. The results reported so far are savings of thousands of dollars in fertilizers, triple the crop yield, faster germination, and an increase in the vigor or survival factors in the plants. This system does not use pyramids to treat seeds, but it was born out of pyramid research. The ECSS is a combination of two things: A. The organic mineral seed coating, B. The electronic seed stimulator.

As a result of the pyramid experiments, other exciting energy devices have come to light. We have found electronic means of generating far more powerful fields than the pyramid, a new pyramid electronic communications device, and several other inventions.

We have found that other shapes such as cones, hemispheres, and even flat planar shapes possess the same properties attributed to the pyramid. Some of the other devices have an advantage over the pyramid in that they are not as critical in alignment to the earth's magnetic field as the pyramid.

For this reason, we have included a section on cones and cone mathematics for the experimenter. We have also found that the pyramid does not even have to have a covering to be effective. The corner outline in the form of aluminum tubes, brass welding rods, wood, and even string can have effects on substances.

We are also including a section on the Experimental Sensor, the small pocketable antenna that acts as a pyramid and is the most powerful small device we have conceived. It is by far more powerful than small pyramid generators, and other items we have conceived in the past.

The following experimental sections include mathematics, measurements, and other pertinent information for the serious experimenter who desires to pursue his own research into these unusual energies. Included is information on pyramids, cones, the sensor, the EES 2000 experimental course, the Flanagan News Letter, the Neurophone patent, and other pertinent information.

Pyramid Construction

Pyramid models can be made out of many different materials.

G. PATRICK FLANAGAN

Any non-ferrous or non-magnetic material is suitable. I have made models out of cardboard, wood, copper, concrete, aluminum, plastic, bakelite, epoxy glue, and glass. We have made pyramids with solid walled construction, and with no walls. The only criteria are that the pyramids must be made according to the mathematics of the Great Pyramid, and they must be free of interference from surrounding water pipes and objects that would interfere with the Earth's magnetic fields. The pyramid model must be properly oriented to the magnetic fields, so the experimenter needs to purchase a compass from a sporting goods store.

In order to control the experiment properly, a cubical box of equal volume may be made out of the same material as the pyramid. It is generally believed that the item to be treated must be placed in the area that the King's Chamber normally occupies in the Great Pyramid. This area is approximately 1/3 of the way up from the base of the pyramid. This point corresponds to the center of gravity of the pyramid, that is, the point of balances.

In a cube, the control substance is placed at the center of gravity of the cube, a point half-way up from the base, or center of the cube. The cube is used as a control, as we normally build our houses on a cubical plan, and all scientific constants refer back to the cube as a basic unit of measure.

In order to build a pyramid, the first pyramid can be made out of cardboard. Posterboard from a drugstore or art supply house is perfect. The edges of the triangles can be taped together with transparent Scotch tape.

You need to make four identical triangles as in the diagram

below:

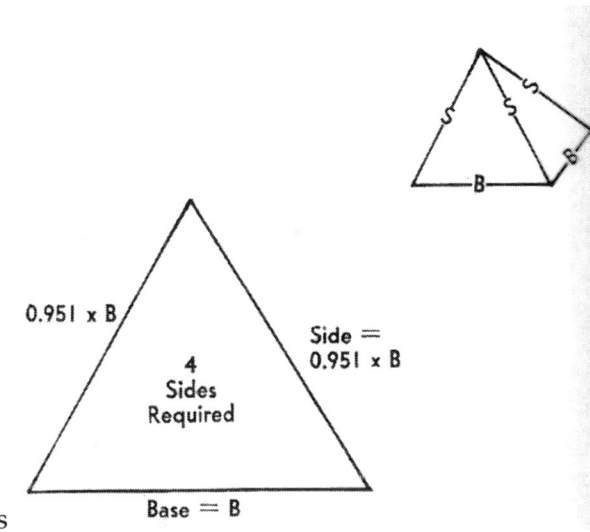

Notice that two sides of the triangle arc shorter than the base. This is in accordance with the mathematics of the Great Pyramid. More details of the math will follow.

Any size can be constructed simply by following the simple formula as outlined. A table of pyramid dimensions is presented below for those who wish to avoid calculation. The dimensions are rounded to the nearest tenth of an inch as normal construction practice will not allow greater accuracy.

BASE	SIDE	HEIGHT	VOLUME	CER*	Side of Equal Vol Cube
6"	5.7"	3.8"	45.6	2.3	3.6"
10"	9.5"	6.4"	213.3	10.7	6.0"
12"	11.4"	7.6"	364.8	18.2	7.2"
20"	19.0"	12.7"	1693.3	84.7	11.9"
24"	22.8"	15.3"	2937.6	146.9	14.3"
36"	34.3"	22.9"	9892.8	494.4	21.5"
4'	3'10"	2'7"	13.7 ft^3	0.7 ft^3	2'5"
6'	5'8.5"	3' 10"	46	2.3	3'7"

"CER—Critical Energy Ratio, expressed in cubic units.

All dimensions in the table are calculated with pi accurate to ten decimal places and rounded to the nearest tenth of an inch.

Once your pyramid is constructed, it must be placed in an area free of the effects of steel building girders, water pipes and steel conduits. Most failures in pyramid experiments are the result of improper pyramid positioning and/or orientation of the magnetic field of the Earth.

With the compass, determine the direction of magnetic north. The pyramid will be aligned according to the diagram below:

BEYOND PYRAMID POWER

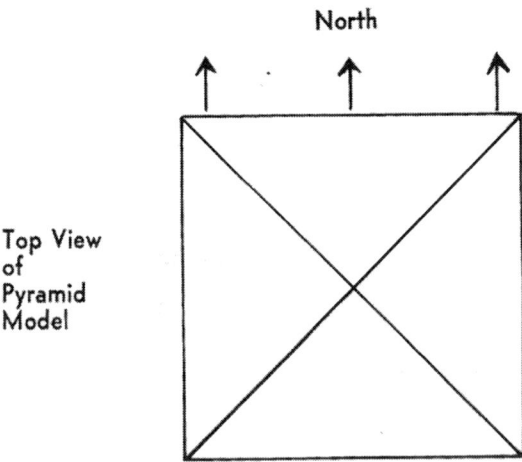

Top View of Pyramid Model

Critical Energy Ratio

Experimentation has shown that there is a Critical Energy Ratio, or GER between the pyramid and the item to be treated. This ratio is based on the volume of the pyramid and the volume of the item to be treated. A pyramid of certain dimension only has a certain capacity. When the capacity of the pyramid is exceeded, the experiment is doomed to failure.

The volume of the item to be treated *must not exceed 5% of the volume of the pyramid.*

The volume of the pyramid can be calculated by the following formula:

$$\text{Volume} = \frac{B^2 H}{3} \qquad B = \text{Base} \qquad H = \text{Height}$$

or is approximately equal to $0.212\ B^3$.

Items to be treated in the pyramid must be prepared in the best ways for the success of the experiments. These will be covered in the section under Mummification and Preservation.

Cube of Equal Volume

The control cube of equal volume is made of five square pieces of cardboard. The base dimension of the side of the cube can be easily calculated by the following formula.

Side length off cube for
equal volume to pyramid $= 0.596 \times$ Base length of pyramid

$$\text{Cube } S = 0.596 \, B.$$

Mummification

An item is mummified when it has dehydrated with a minimum amount of bacterial decay. Small samples of just about any foodstuff can be dehydrated in the pyramid, showing little or no deterioration. It is best that the item to be mummified be exposed to as much air us possible. Therefore, holes or windows should be cut into the side of the pyramid so air can get to it. The item to be mummified should be shredded or diced to allow maximum surface area to be exposed to energy and air. Some specimens such as birds, small fish, and insects can be placed whole in the pyramid and left for an extended period of time. Specimens can include meat (various types), cheese, fruits, insects, tropical fish, vegetables, and anything else the reader may desire to try.

BEYOND PYRAMID POWER

An excellent idea is to construct a pyramid food dehydrator to preserve foods for later eating. A small heater fan to circulate air through the pyramid would be ideal for this purpose. The temperature and air flow should be adjusted so it never exceeds 105 degrees in order to preserve enzymes. The flavor of pyramid dehydrated fruit must be tasted to be believed! I don't know why someone hasn't tried to capitalize on this by making pyramid dehydrated fruits for health food stores. (I get a royalty if you do).

Preservation

The pyramid model will preserve food items in much the same way as a refrigerator that is it will extend the useful life of a substance three or four times its normal life span. In the case of preservation of food, it is best to leave the item whole in the pyramid until you desire to eat it. A number of pyramid researchers treat their groceries in the pyramid for several hours before putting them away.

People and Pyramids

As reported in Pyramid Power, the pyramid increases the alpha activity in subjects when they are in a pyramid. These tests have been performed on a number of people at several different universities with the same results. When a pyramid was lowered over the head of a person connected to an EEG machine, the alpha rhythm activity of the brain dramatically increased within one minute.

I have received many letters from pyramid fans reporting how the pyramid has improved sexual sensitivity and increased pleasure of the sexual act. In one particular case, I built a pyramid water bed for a millionaire. The cost in parts alone was well over $5,000. I later heard that he spent three months in his bedroom and never left his bed.

Measurements of the Great Pyramid

The following are a few measurements of the Great Pyramid of Giza.

SOME MEASUREMENTS OF THE GREAT PYRAMID

Height to imaginary topstone's apex: ....5,813"
Base measurement of one face: ....9131.06"
Actual height, consequence of destruction ....5,496"
Edge (face) Corner to Apex: ....8,687.88"
Apothem center of base face to apex............................,....................7,391.56"
Base angle, base to level ground ....51° 51' 14.3"
Apex interfacial angle: .... 76° 17' 32"
Edge to level ground: .... 41° 59' 50"
Dihedral angle, face to face edge: ....112° 25' 38.88"
Apex, edge to edge: ....96°
Circuit of the base .... 36,524.24"
One terrestrial year equals 365.24 days!
Sum of base diagonals ....25,826.5"
Precession of our equinox is 25,827 years!
Volume of Pyramid .... 90,000,000 ft³
Area covered ......,.......................... 54,000 square meters (13+ Acres)
Weight..5,955,000 TONS

Difference between English and Pyramid Inch 0.0011"

BEYOND PYRAMID POWER

The number five appears throughout the Pyramid. Five sides, five corners. The temperature inside the King's Chamber is 68° Fahrenheit. This is exactly one fifth the distance mercury is raised in a tube between freezing and the boiling point of water at Sea Level. This is also the optimum temperature for health and long life.

One half the base of the Pyramid divided into the apothem is equal to <l> For example:

9131/2:4565.5;7387.15/4565.5:1.6180374

Compared to the base, the edge length of the face of the Pyramid is 4.9% less: 8683.58/9131:O.9509998.

The height of the Pyramid is 36.4% less than the base: base: 9131, height=5813; 5813/9131; 0.6366224.

Relation of golden section in Great Pyramid

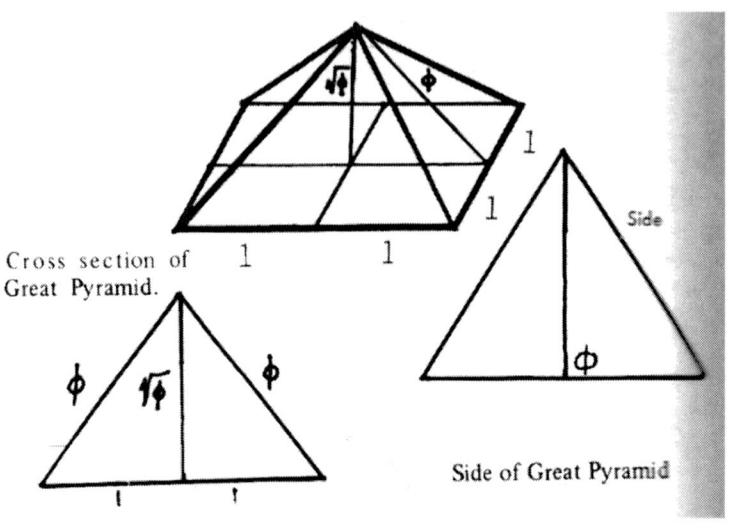

Pyramid Mathematics

If the height of the pyramid is doubled and divided into the perimeter of the base of the pyramid, the result is pi, the unending constant of mathematics used to determine the ratio of the diameter of a circle to the circumference of a circle.

It is said that the pyramid squares the circle! There are numerous references to pi in mathematics literature. Around the turn of the century, there were arguments that any physical constant in the universe must have a finite value. Buckminster Fuller writes that pi must have a finite measurement for it would be impossible to make a perfect circle if this were not so. The arguments about *pi* go on and on. Pi is used quite successfully in present mathematics to determine the orbits of spacecraft so it will do for now.

Pi = 3.1415926535897932384627433832795028841971 6-

BEYOND PYRAMID POWER

9399375105820974+ . . . . .

For our purposes, we use pi rounded to the ninth decimal place, pi : 3.141592654

Pyramid ratios according to Pi are as follows:

$$\text{Perimeter} = 2H\pi$$

$$\text{Base side} = \frac{\text{Perimeter}}{4}$$

$$H = \frac{2B}{\pi}$$

The Golden Ratio or PHI φ

The Great Pyramid is the only structure known that incorporates the ratios of PI and PHI into one building. This makes it all the more unique as the ratio of PHI and PI are not supposed to have been known when the pyramid is supposed to have been constructed.

PHI is the magical, mystical unending number 1.618033-989+.......... This number so fascinated the artists during the Renaissance that they spent most of their time exploring its endless possibilities in utilizing its ratios in their paintings and sculptings. Modern architects use it in building design as it has been found to be the most pleasing ratio. At one point in history, the ratio for PHI was so carefully guarded that to reveal it to a non-
initiate was a crime punishable by a death of slow torture.

Why is PHI so important? It is the mystical number that is present in the design of all living organisms. It seems to recur over and over in all forms of life. The logarithmic spiral found in the Chambered Nautilus, the shape of the growing fetus, the ancient Yin-yang symbol of life from the orient, the daisy, the sunflower, the pine cone, pineapple, elephants' tusks, the human body, canaries' claws, and the human ear are but a few examples of organic structures that incorporate PHI.

PHI is closely associated with the mathematics of life. It has been proposed that all living structures are closely related to the invisible etheric field surrounding them. This aura is said to shape all life forms. It is interesting then that PHI may be a mirror of the basic mathematics of life energy. It is no wonder that the pyramid shape incorporates PHI into its structure. The pyramid greatly affects life energy processes

The Fibonacci Series

There is a number system used in botany to determine the basic plant leaf distribution in trees and other plants. This series of numbers is called the Fibonacci Series after its discoverer. This series is unique in that by its use, the value of PHI can be determined.

Mathematics texts state a formula for PHI as

$$\text{PHI} = \frac{1/2 + \sqrt{5}}{2}$$

Calculated on this basis, PHI: 1.618033989 to nine decimals. We use this value in evaluating the Fibonacci series that follow.

The Fibonacci series of numbers is determined in the following

BEYOND PYRAMID POWER

manner: Each succeeding number is determined by the sum of the two preceding numbers, beginning with the lowest whole number. Thus: 1, 2, 3, 5, 8, 13, 21, 34, 55, 89, 144, 233, 377, 610, 987, 1597, 2584..... are all succeeding terms in the series. I will explain later how this series is used to determine the distribution of leaves, pine cone, pineapple buds, and sunflower seeds.

If any number is divided into its successor, we get a ratio that approaches PHI as the numbers get higher and higher:

#	Ratio		Percentage Error To ϕ
1	2/1	= 2	+23.6%
2	3/2	= 1.5	7.3%
3	5/3	= 1.6̂66+ ...	3.0%
8	8/5	= 1.6	1.1%
13	13/8	= 1.625	0.43%
21	21/13	= 1.6153846	0.1637%
34	34/21	= 1.6190476	0.06%
55	55/34	= 1.6176471	0.025%
89	89/55	= 1.6181818	0.009%
144	144/89	= 1.6179775	0.0034%
233	233/144	= 1.6180555	0.0013%
377	377/233	= 1.6180257	0.0005%
610	610/377	= 1.6180371	0.0002%
987	987/610	= 1.6180327	0.00007%

The last ratio is accurate to the value of PHI by 70/1,000,000ths of 1%

If we examine any leafy plant, the distribution of kernels on a pine cone, the distribution of buds on a pineapple, we find there are two things occurring: the leaves, kernels, buds, etc., all rotate around a central axis in clockwise and counter clockwise directions; and the number of leaves etc., in one direction is always different from the distribution in the other direction. This difference is always according to the Fibonacci number system! For example, in one pine cone examined, there were 21 kernels in a clockwise direction and 34 on the counter clockwise direction. In a sunflower examined, there were 21 kernels in a clockwise direction and 55 in the counter direction. These ratios are also incorporated as a PHI ratio in the dimensions of living things. For example in the body of man:

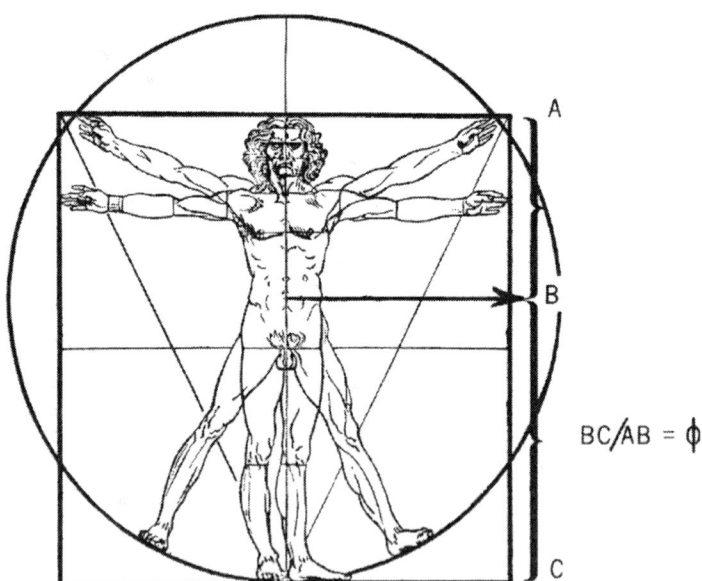

The above study in human proportion by Leonardo da Vinci shows the Golden Section as a basis of measurement on the human body. This sacred proportion appears throughout nature, and is the governing ratio in the Great Pyramid.

The famous painting of the Last Supper by DaVinci is based on the PHI ratio. PHI is also known as the SACRED CUT. This is

BEYOND PYRAMID POWER

represented by the fact that it is the exact ratio in which we can divide a line AC by B in such a manner that AC/AB: 1.618 . . ., is the same as AB/BC: 1.618 . . . This magic ratio was used in the Renaissance by all the great masters and is considered to be the most aesthetic proportion. The body of man is divided by this ratio. The sacred five pointed star's diagonals divide each other by this ratio. These proportions occur throughout nature.

A ———————————————— B ———————— C

$$\phi^2 = \phi + 1 \qquad 1.618033989^2 = 2.6180339 \times 1.2 =$$

$$\pi = \phi^2 \times 6/5 \qquad 3.141640788$$

TRUE
$$\pi = 3.141592654 \qquad \% \text{ error} = 0.0015\%$$

The Pi and Phi factors in the mathematics of the Great Pyramid are substantiated by the following facts. The original alabaster covering on the pyramid has been destroyed by vandals. The result of this destruction is that accurate measurements cannot be made on the present pyramid. However, a casing stone was found at the base of the pyramid and the angle of this stone was measured. It was found that its base angle was 51° 51′ 14.3″. This angle has been called the PI angle because of the following relationship:

$$\text{Arc Tan } \frac{4}{\pi} = 51° 51' 14.3''$$

The arris angle of the pyramid, the angle formed by the ground and the rising edge of the pyramid is:

$$\text{Arris Angle} = \frac{(\sqrt{2})^3}{\pi}$$
$$= 41° \, 59' \, 50''$$

We can see that the PI ratio plays a significant part in the pyramid architecture.

If we examine the face or side triangles of the pyramid, further validation of the part played by PHI, becomes apparent:

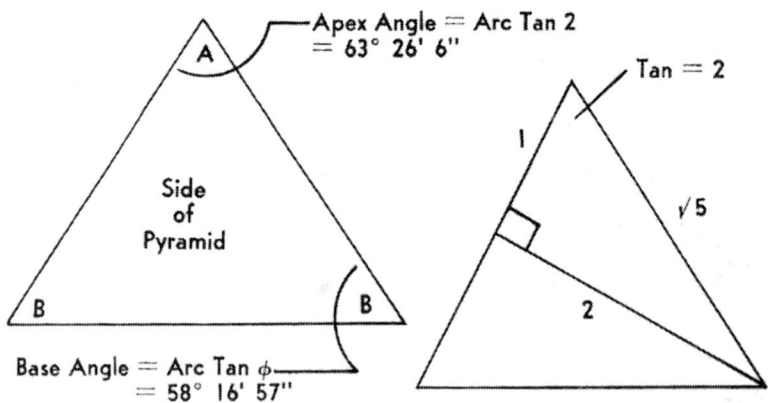

We can see from the above diagrams, that the angle of the apex of the face of the pyramid is equal to the Arc Tangent of 2, or 63° 26' 6". The base angles of the pyramid face are then 58° 16' 57" or the Arc Tangent of PHI.

Another interesting feature that can be used in the accurate construction of a pyramid is the use of the fact that the apex

BEYOND PYRAMID POWER

angle is the arc tan of 2.

A very accurate pyramid can be constructed by means of a ruler and drawing compass. If one wishes to construct an accurate pyramid out of wooden panels, the edges of the triangles must be accurately beveled in order to get a good pyramid. The bevel angles are shown below.

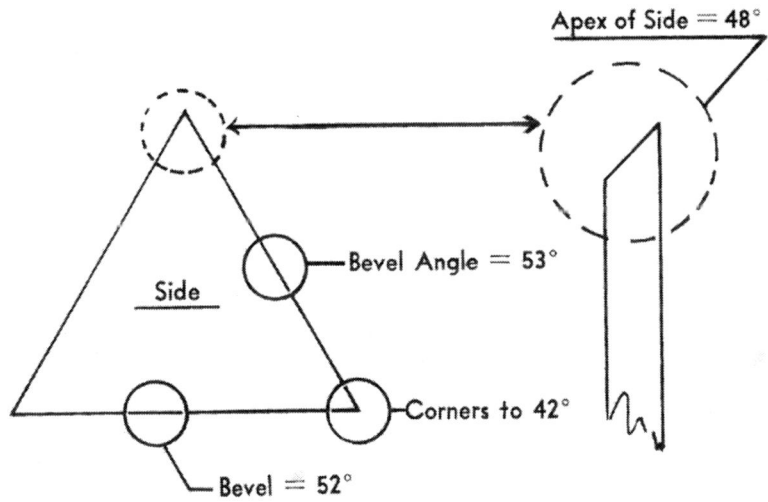

This can be performed by a good woodworking shop.

The above beveling can be performed on rectangular wood poles for construction of an open framed pyramid.

ALUMINUM TUBE PYRAMID

A very good frame is an open faced aluminum frame meditation pyramid. I made my own unit in this manner. The four six foot aluminum tubes are hinged by a unique method as shown below:

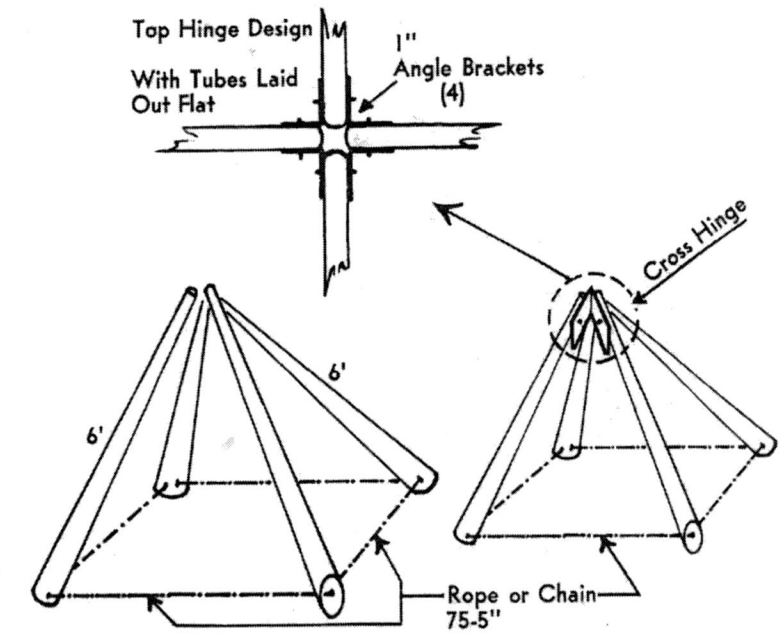

The angle brackets for the hinge can be purchased from any hardware store. The brackets can be fastened by means of nuts and bolts, or pop rivets. The chains connecting the tube members can be of brass or even nylon cord will do. The lengths are calculated to give the proper pyramid angle.

I have given my own dimensions for the reader to easily follow. This pyramid can be constructed for less than ten dollars! I have a ten foot base one over my king sized bed!

BEYOND PYRAMID POWER

This ends the section on Pyramids. I mentioned that cones are as good as pyramids and the same experiments can be performed with cones. The following section is on cone construction and mathematics.

CONE CONSTRUCTION

Belizal, Cameron, and others found that the cone worked just as well and even better than a pyramid. I have found this to be true. On examining the cone, it became obvious why this is so. A pyramid is any shape that is constructed of triangular sides. The simplest pyramid is a tetrahedron or three sided pyramid. If we start adding sides to the pyramid, it goes through a series of transformations. A pyramid of 1,000 sides no longer looks like a pyramid, but resembles a smooth sided cone. A cone is therefore a pyramid with an infinite number of sides. The unique feature of the cone is that it is always aligned to the magnetic fields of the environment no matter where it is placed.

I have made cone generators in the same way the old pyramid energy generators or PEGs were constructed and have had very exciting results from their use. The cone generator makes use of energy radiated from the tops of the cones in the same way the original PEG made use of energy radiated from the tops of the pyramids.

A Kirlian photograph of the PEG and its radiation is included for the reader.

KIRLIAN PHOTO OF PYRAMID
ENERGY IN ACTION

A drawing of a cone generator is shown below:

The cone generator is arranged in a circular or hexagonal configuration as shown.

Cones can be made out of the same materials described in the pyramid section of this paper. They are most easily made from cardboard or paper. I have made many out of aluminum laminated paper. A cone is made by first cutting out a circle. A pie shaped section is then marked off on the circle. A cut is made along one of the lines to the center, and the sections are made to overlap.

BEYOND PYRAMID POWER

As the overlapping occurs, a cone is formed, and the pie shaped section is folded into the center of the cone. The mathematics of cone making are outlined below:

$$P = 360 - (360 - \cos A)$$
$$X = \frac{R}{\cos A}$$

Where: P = angle of the pie section removed from the circle

A = base angle of cone to ground
X = the radius of the circle
R = radius of the completed cone

The P angles of the pie shaped sections are calculated below for several base angles of cones. The base angle equal to the Great Pyramid is a good one to try. It is a Great Pyramid with an infinite number of sides.

Base Angle (A)	P (pie section)
25°	33.7°
45°	105.4°
Pyramid Angles 50° 51' 14.3"	137.369°
60°	180°

I have made cone meditation tents and have had very good success with them in fact I prefer the cone to the pyramid. A very simple cone meditation tent is made by using one upright pole stuck in the ground, and stringing wires or nylon cords from the top of the pole in a circle. Maybe the Indians knew what they were doing with their Tipis!

The volume of a cone can be calculated as below:

$$\text{Vol Cone} = \frac{\pi R^2 H}{3} \qquad \begin{array}{l} H = \text{height} \\ R = \text{radius of cone base} \end{array}$$

The Critical Energy Ratio applies to cones as it does to pyramids.

I have found that other shapes such as spheres, and hemispheres also have energy characteristics and are fun to experiment with.

BEYOND PYRAMID POWER

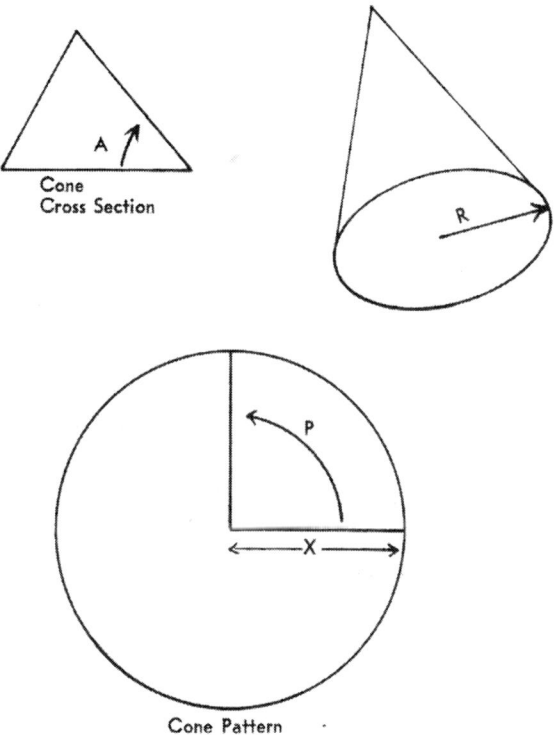

The Flanagan Experimental Sensor

The Sensor is basically a flat pyramid of an infinite number of sides. It is a flat disc with ten rings of gold plated copper. As it is only three inches in diameter, it may be carried in the pocket or worn as an attractive necklace when suspended on a chain.

All experiments that can be performed with a pyramid can also be performed with the Sensor. It emits the same energy field as a pyramid, with the lightweight compact advantage of being able to fit into a pocket for easy portability.

In the fields of microwave and higher frequency energy, it is common practice to modify three dimensional antennas to fit into two dimensional or flat planes. This practice is very common in areas where compactness is desired.

In my work with pyramids, and then cones, it became obvious that I could apply the same technology to the cone in order to make the cone-pyramid a portable, easily carried experimental tool. My earlier attempts to miniaturize the pyramid with my PEG's or Pyramid Energy Generators were very successful. However, the PEG's were still very large, heavy, and cumbersome. In my search for a better answer, the Experimental Sensor was developed.

It is called a Sensor as it is very sensitive to biocosmic energy from the solar system. It has a double advantage in that it can be used in any position without the need of alignment to the Earth's fields. The old generators were very heavy as they had to have their own magnetic fields built in.

We have had very exciting reports back from Sensor users. It makes potted plants thrive; seeds treated germinate faster; water treated is excellent for shampooing, and making coffee and tea; instant coffee loses its bitterness; pipe and cigarette tobacco become mild and lose their harshness and offensive odor; tropical fish live three times as long when their water is treated; foods such as meat, fruits, and vegetables taste better after treatment; swimming pools don't build up algae deposits even when no chemicals are added to the water; in fact, there are some farmers in Texas who put the Sensor in their watering tanks to keep algae from building!

The Sensor has also been shown to create energy balance in acupuncture meridians when it is worn over the meridians in the body. These changes in the meridians can be measured by various means.

The following description of one test is an example of the Sensor's effect on the energy in the acupuncture system of the body.

The Sensor and Acupuncture Energy Balance

When the Sensor is worn as a necklace, or placed on the forehead, it has the effect of recharging the Vessel of Conception, the acupuncture meridian that is a storage depot for bodily energy reserves according to the Chinese.

A simple experiment to demonstrate this can be performed by anyone. You need a friend to work with. One thing reported in acupuncture is that the strength of the muscles of the body is affected by the amount of etheric energy or Chi in the acupuncture meridians. One way of testing the strength is by testing the pull of one muscle group.

One of the easiest to test is the pectoralis major clavicular muscle. These muscles are isolated by having the subject stand with one arm held out in front, elbow straight, with the hand open and thumb pointed toward the feet. Pressure is gradually exerted on the top of the wrist to force the arm down toward the feet. The elbow must remain straight at all times. The subject resists the pressure by pushing upward. By judging the amount of resistance, the amount of strength can be determined.

By purposely weakening the energy in the central meridian, this group of muscles and the whole body can be considerably weakened. This is done by running the finger tips in a downward direction from the chin to the crotch area in rapid strokes. When the hand reaches bottom, it is removed from the body, brought up to the chit for another downward stroke. At the end of 10 or 20 strokes, test the strength of the muscles again. They should be weakened considerably.

If we then place the Sensor lightly against the middle of the forehead for one minute, the strength is not only restored, but it is actually increased! This experiment demonstrates the relationship of biocosmic energy to the energy forces in living organisms. This is why this energy can have so many effects on organic substances.

The previous experiments were merely suggest ions. The reader will think of many more for his own experimentation.

We are publishing a newsletter, The Flanagan Newsletter. This is a Bi-Monthly news publication and contains articles on biocosmic energy, cones, pyramids, acupuncture, yoga, and many other interesting subjects. We are urging our readers to write in their experimental results so these can be published and shared with our readers.

BEYOND PYRAMID POWER

For information, write to:
Dr. G. Patrick Flanagan
271 S Airpark Rd,
Cottonwood, AZ 86326
Help@PhiSciences.com

CPSIA information can be obtained
at www.ICGtesting.com
Printed in the USA
LVOW10s2229200317
527885LV00012B/692/P

9 781530 859153